Constellations

by Grace Hansen

Abdo Kids Jumbo is an Imprint of Abdo Kids
abdobooks.com

abdobooks.com

Published by Abdo Kids, a division of ABDO, P.O. Box 398166, Minneapolis, Minnesota 55439.

Abdo Kids Jumbo™ is a trademark and logo of Abdo Kids.

Printed in the United States of America, North Mankato, Minnesota.

102019

012020

Photo Credits: Alamy, Getty Images, iStock, Library of Congress, Shutterstock

Production Contributors: Teddy Borth, Jennie Forsberg, Grace Hansen
Design Contributors: Dorothy Toth, Pakou Moua

Library of Congress Control Number: 2019941292

Publisher's Cataloging-in-Publication Data

Names: Hansen, Grace, author.

Title: Constellations / by Grace Hansen

Description: Minneapolis, Minnesota : Abdo Kids, 2020 | Series: Sky lights | Includes online resources and index.

Identifiers: ISBN 9781532189067 (lib. bdg.) | ISBN 9781532189555 (ebook) | ISBN 9781098200534 (Read-to-Me ebook)

Subjects: LCSH: Constellations--Juvenile literature. | Stars--Juvenile literature. | Space--Juvenile literature. | Astronomy--Juvenile literature. | Star gazing- Juvenile literature.

Classification: DDC 523.8--dc23

Table of Contents

What Is a Constellation?

A constellation is a **cluster** of stars. These clusters can make shapes that we recognize.

There are 88 named constellations. Many of them were named by the **Ancient Greeks**.

Claudius Ptolemy was an **astronomer**. He named 48 constellations.

One named by Ptolemy is Ursa Major. It is also called the Great Bear. Seven of its brightest stars make up the Big Dipper.

Big Dipper

Sirius is the brightest star in the night sky. It is easy to spot! It is a part of the Canis Major constellation.

sirius
sirius

Spotting a Constellation

Being able to spot constellations depends on where you live. It also depends on when you're looking.

The Earth **orbits** the sun. It spins as it orbits. The equator splits Earth into two halves. These halves are called hemispheres.

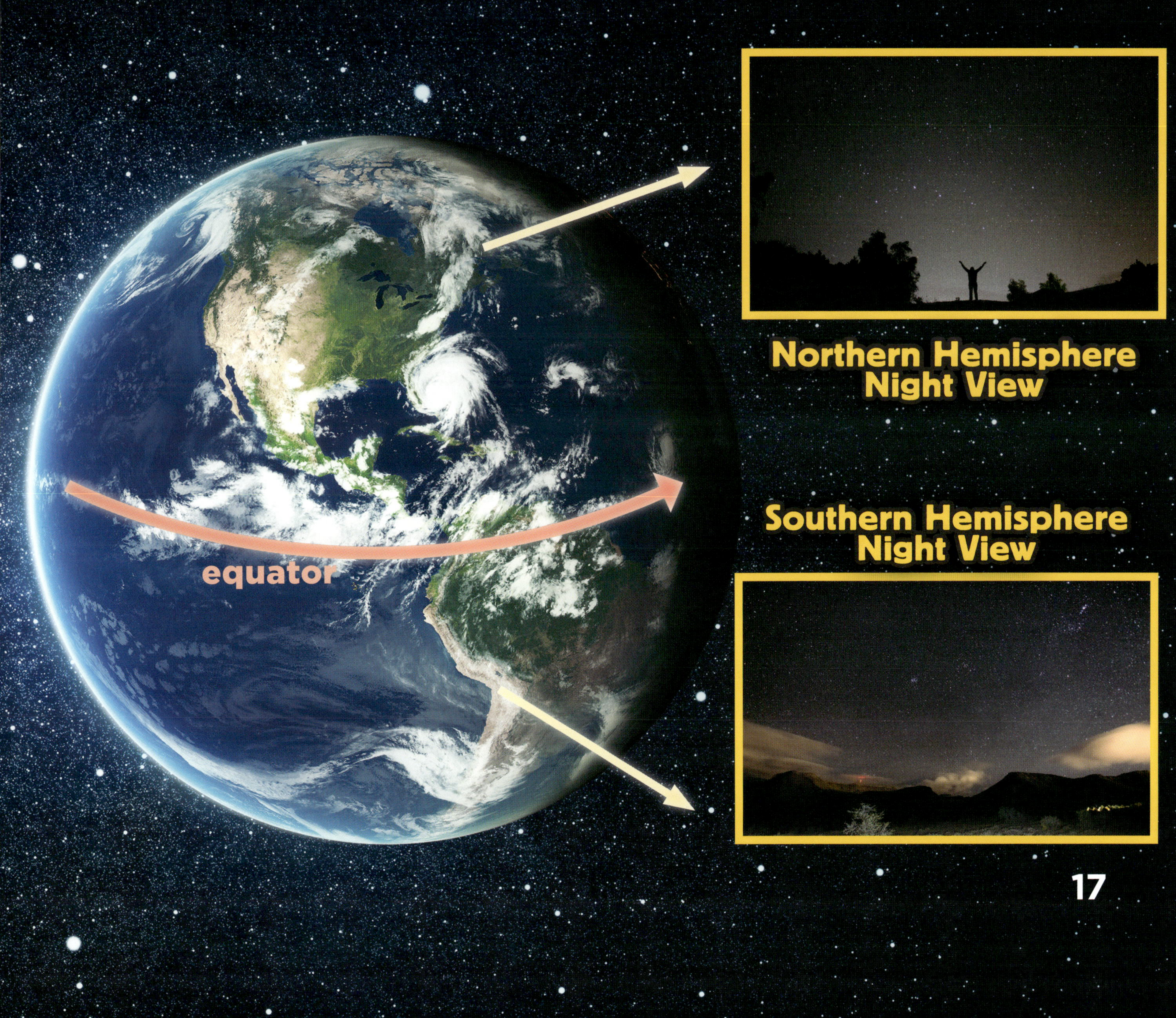

equator

Northern Hemisphere Night View

Southern Hemisphere Night View

People who live in the Northern Hemisphere can see Canis Major in winter. Those who live in the Southern Hemisphere may be able to see it in summer.

Canis Major
Sirius

The Zodiac

The **zodiac** is made up of 13 constellations. The zodiac star groups make a circle around Earth. As the Earth spins, different zodiac constellations can be seen.

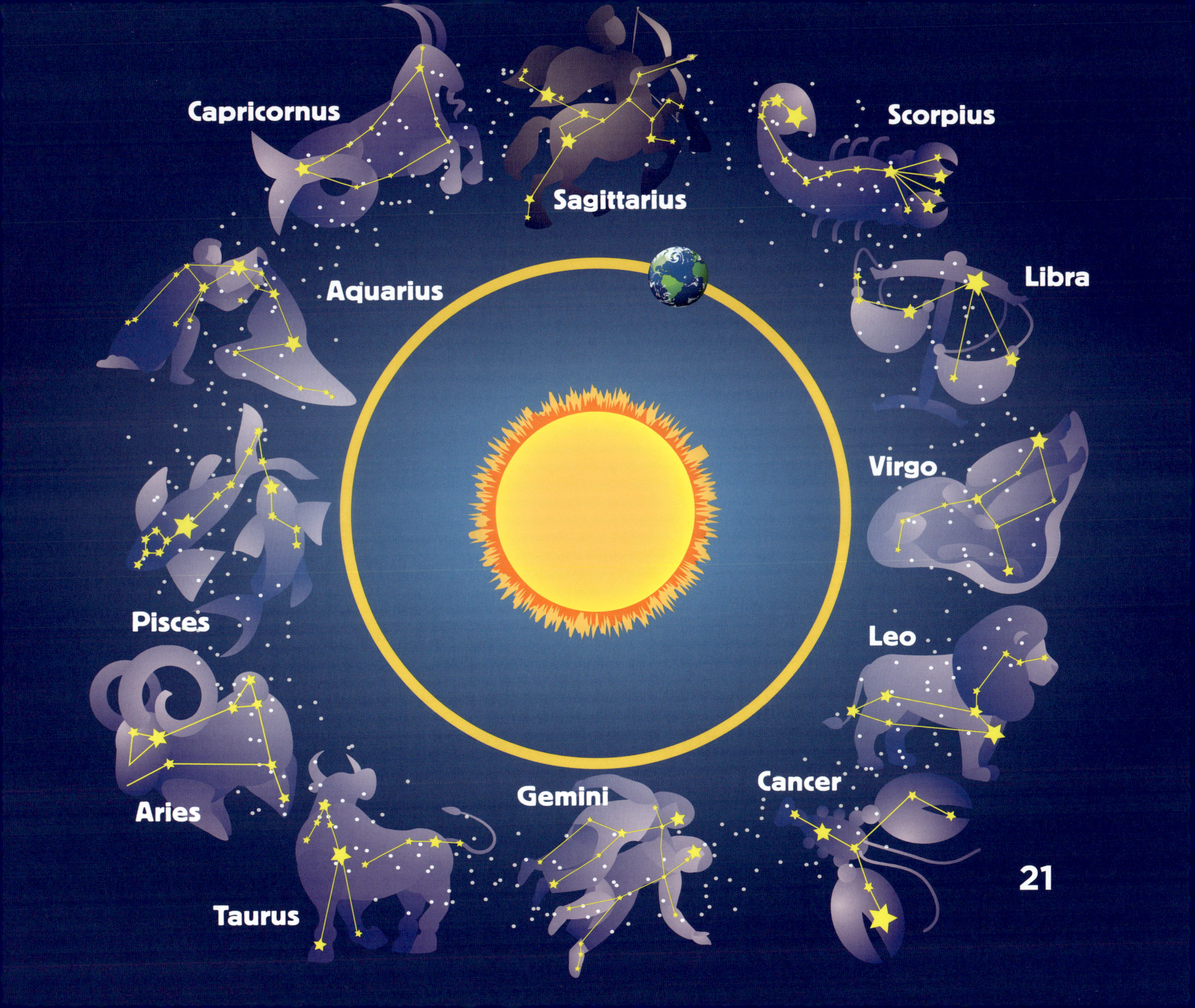

Capricornus

Sagittarius

Scorpius

Aquarius

Libra

Virgo

Pisces

Leo

Cancer

Gemini

Aries

Taurus

More Facts

- The word constellation comes from the Latin term meaning “set with stars.”
- There are 12 constellations used for the **zodiac** calendar. They are Capricornus, Aquarius, Pisces, Aries, Taurus, Gemini, Cancer, Leo, Virgo, Libra, Scorpius, and Sagittarius.
- The 13th zodiac constellation is Ophiuchus. It was left out of the group on purpose. It is thought that scientists wanted to divide the sun’s 360 degree path by 12 and not 13.

Glossary

Ancient Greeks – people who lived in Greece around 2,500 years ago. The Ancient Greeks were great thinkers, athletes, artists, and more.

astronomer – a scientist who studies the universe beyond the earth.

cluster – group.

orbit – a curved path in which a planet or other space body moves in a circle around another body.

zodiac – an imaginary belt in space that includes the paths of the sun, moon, and planets as seen from Earth. The zodiac is divided into 12 equal signs or parts. Each part is named for a constellation that appears within the belt.

Index

Visit **abdokids.com** to access crafts, games, videos, and more!

Use Abdo Kids code

SCK9067

or scan this QR code!